The Open University

Science: A Second Level Course

COMPARATIVE PHYSIOLOGY

Introduction—The Animal Kingdom

Prepared by the Course Team

The Open University Press

THE COMPARATIVE PHYSIOLOGY COURSE TEAM

Chairman and General Editor

M. E. Varley

Unit Authors

D. Adamson

H. Adamson (*Churchill Fellow 1971*)

R. M. Holmes

S. W. Hurry

J. N. Thomas

M. E. Varley

Editor

F. Aprahamian

Other Members

E. A. Bowers (*Staff Tutor*)

N. R. Chalmers

B. Cordell (*Staff Tutor*)

R. D. Harding (*Course Assistant*)

S. W. Hurry

T. Laryea (*BBC*)

J. McCloy (*BBC*)

G. D. Moss (*IET*)

S. P. R. Rose

The Open University Press,
Walton Hall, Milton Keynes.
MK7 6AA.

First published 1971   Reprinted (with corrections) 1973 and 1977

Designed by the Media Development Group of the Open University.

Printed in Great Britain by
Billing & Sons Limited, Guildford, London and Worcester.

ISBN 0 335 02291 X

This text forms part of an Open University Second Level Course. The complete list of units in the course appears at the end of this text.

For general availability of supporting material referred to in this text, please write to Open University Educational Enterprises Limited, 12 Cofferidge Close, Stony Stratford, Milton Keynes, MK11 1BY, Great Britain.

Further information on Open University courses may be obtained from the Admissions Office, The Open University, P.O. 48, Milton Keynes, MK7 6AB.

Contents

Table A	List of Scientific Terms, Concepts and Principles	4
	Objectives	5
	Study Guide	6
1.0	Introduction	7
1.1	Principles of Taxonomy	8
1.1.1	Naming organisms	8
1.1.2	Some complications	9
1.1.3	Conclusion	10
1.2	Survey of Living Organisms	10
1.3	The Animal Kingdom (excluding Protozoa)	12
1.3.1	Cnidarians	13
1.3.2	Worms	16
1.3.3	Molluscs	21
1.3.4	Echinoderms	23
1.3.5	Arthropods and vertebrates	24
1.4	Phylogeny—the Interrelationships between Phyla of Metazoan Animals	26
1.5	Summary	32
	References	32
	Glossary	32
	Self-assessment Questions	33
	Self-assessment Answers and Comments	36
	Answers to In-text Questions	37

Table A

List of Scientific Terms, Concepts and Principles

Taken as prerequisites			Introduced in this Unit			
1	**2**	**S100 Unit No.**	**3**	**Page No.**	**4**	**Unit No.**
Assumed from general knowledge	Introduced in a previous Unit		Developed in this Unit or in its set book(s)		Developed in a Later Unit	
	species	19	family	9		
	genus	19	order	9		
	virus	17	class	9		
	bacteria	17, 18	phylum (animal)	9		
	autotrophe	20	Division (plants)	9	Division (plants)	2
	heterotrophe	20	Procaryota	10		
	Protista	18	Eucaryota	10		
	Fungi	18	Animal Kingdom	11		
	muscle	18	Plant Kingdom	11	Plant Kingdom	2
	segment	21	Cnidaria (coelenterates)	13		
	blood-vascular system	18	enteron	13		
	exoskeleton	21	mesogloea	13		
	endoskeleton	21	types of symmetry	15		
	adaptive radiation	21	levels of organization	28		
	convergence	21	coelom	17		
	parallel evolution	21	pseudocoel	17		
	mutation	19	haemocoel	21		
	natural selection	19	Platyhelminthes (flatworms)	16		
	niche	20	Aschelminthes (nematode worms)	17		
	herbivore	20				
	carnivore	20	Annelida	17		
	parasite	20	Echinodermata	23		
	cilia	18	musculoepithelial cells	13		
	Vertebrata	21	sessile, sedentary and motile habits	14		
	Mollusca	21	super-phyla	26		
	Arthropoda	21	phylogeny	26		
	fossils	21	Hemichordata (*Cephalodiscus*)	20		
			Chordata	9, 30		

Objectives

After you have studied this Unit, you should be able:

1 To demonstrate knowledge of items in Table A by:

(a) defining them in your own words;
(b) recognizing definitions or applications of them;
(c) relating them with given statements.

(Tested in *SAQ* 1.)

By selecting from a matrix or multiple choice array:

2 To contrast features of structure and life history associated with:

(a) aquatic and terrestrial habits in animals;
(b) sessile, sedentary and mobile habits in animals;

(Tested in *SAQ*s 3 and 4.)

3 To contrast the types of activity associated with:

(a) cilia; (b) musculoepithelial cells; (c) an unsegmented acoelomate body; (d) an unsegmented body with a coelom or a haemocoel; (e) a segmented coelomate body; (f) an exoskeleton; (g) a bony endoskeleton.

(Tested in *ITQ*s.)

4 To contrast the basic features of Procaryota and Eucaryota.

(Tested in *SAQ* 2.)

5 To compare and contrast the basic features of:

Cnidaria; Platyhelminthes (flatworms only); Aschelminthes (nematodes only); Mollusca; Arthropoda; Echinodermata; Hemichordata (*Cephalodiscus* only); Chordata (vertebrates only).

(Tested in *SAQ* 5 and *ITQ*s.)

6 To recognize examples of invertebrate animals that illustrate adaptive radiation and convergence.

(Tested in *SAQ* 6.)

Study Guide

The Unit starts by defining the scope and nature of Comparative Physiology as a discipline (Section 1.0). Because it is essentially a comparative discipline it is frequently necessary to refer to named organisms. The naming and description of species is briefly outlined in Section 1.1. The next Section (1.2) deals with the major divisions of the living world and is a preliminary to both Section 1.3 in this Unit and Section 2.2 in Unit 2.

Types of muscle arrangement, of body cavities and of habits of life are described in Section 1.3 for eight animal phyla that illustrate different levels of complexity of structure. In Section 1.4 the evolutionary history of the many-celled animals is discussed in the light of the information given in 1.3. An *Invertebrate Survey* is part of this Course. Sent with this Unit are several pages of the Survey. Future pages will be sent with the appropriate Units. All references to the Survey are indicated: *IS*. In parallel with the Survey you may be referred to *Man and the Vertebrates* by A. S. Romer. This is a set book for the Course. References to it will be shown as *M & V*.

Study Sequence

1 Before starting the Course you should have studied the following S100* Units: 14, 15, 16, 17, 18, 19, 20, 21.

2 For this Unit you will need: the Main Text; the *Invertebrate Survey*; ruler; pencil; rubber; Filmstrips (21a and 21b) from S100, Unit 21; S100, Units 18 and 21.

You can *omit* the S100 Units and filmstrips, if you feel confident about the material in them that is referred to in the Main Text.

You can *omit* Section 1.1 of the Main Text, if you feel confident that you understand the criteria involved in naming species and the hierarchy of taxa above the species level. This information was given in S100.[1]**

---

* *The Open University (1971)* Science: A Foundation Course, *The Open University Press.*

** *The list of references corresponding to superior numbers ([1,2,3] etc.) will be found on p. 32.*

1.0 Introduction

The comparative study of how living organisms function (comparative physiology) must be based on some knowledge of the *structure* of organisms since, as stated in the Guide to the Course, 'form is the physical expression of function'. It also requires as a basis some knowledge of the *diversity* of organisms. This first Unit deals briefly with the major types of organisms and some of the problems of classifying them. It then concentrates on 'invertebrate animals'; these include types which some of you may have had no opportunity of observing as well as the insects and molluscs mentioned in S100, Unit 21. You are assumed to have a sufficient knowledge of the probable relationships among the vertebrate animals from S100, Unit 21 but you can supplement this by reading *Man and the Vertebrates*. The diversity and some of the structure of the Plant Kingdom is covered in Unit 2. The first part of that Unit and the present Unit are thus introductory to the main themes of the Course; but it is essential that you should be familiar with the general habits of life of a range of organisms before you study functions such as the acquisition and utilization of raw materials.

Two hundred years ago it was generally accepted that all living species of organisms had been separately created, each designed for the position that it occupied in the Scale of Nature. As you read in S100, Units 19 and 21, just over 100 years ago the theory of evolution by natural selection replaced the dogma of separate creation by the presently accepted view that all living organisms have evolved from ancestors that were unlike themselves and were probably descended from one or a few common ancestral types that formed as or after the Earth cooled down some 4.5×10^9 years ago. As time passed, through variation and selection the variety of living forms increased: groups of organisms evolved, showed adaptive radiation (defined in S100, Unit 21), some became extinct, others underwent further evolution and others survived unchanged.

The intimate connection between structure and function was illustrated in S100, Unit 21 from the convergent evolution of unrelated or distantly related organisms living in similar habitats or with similar habits—flying and swimming vertebrates were quoted as examples.

Since the present anatomy, physiology and biochemistry of organisms is the result of evolution from ancestral types, study of 'phylogeny' (lines of descent of groups) may help to explain why some organisms have certain structures that are different from those of other organisms and why some function in one way and others in different ways even when living in the same habitat. This is a further good reason for including a survey of living organisms and their interrelationships in a course of comparative physiology. In the final Unit, we shall return to the theme of evolutionary relationships to draw together our studies of the evolution of physiological functions of organisms.

Species are evolving today, as they did in the past. This creates certain problems of nomenclature and definition. Section 1.1 provides you with a synopsis of taxonomic terms and introduces you to the principles, methods and difficulties of biologists who work in museums and other institutions to provide the essential framework within which we can arrange our physiological observations to attain an overall view of living organisms.

There is no set text for this Unit, but you may find it useful to look through the sheets of the *Invertebrate Survey* (*IS*) that have been sent to you.

1.1 Principles of Taxonomy

Study Comment

The most important ideas to learn in this section are:

1 that all living organisms are given a two-part name,
2 the implications of the generic and specific names.

It is desirable but not essential that you should appreciate both the philosophical background of taxonomic studies and some of the problems in this area of specialized study.

If you are pressed for time concentrate on Section 1.1.1.

1.1.1 Naming organisms

It would obviously be highly inconvenient for everybody if there were no way of telling other people which animal or plant one was referring to—assuming the information being communicated was interesting in the first place. To communicate the information accurately and quickly requires that there should be an agreed name for the organism and that there should be agreement about the description of the organism. There have been cases of confusion caused by the same name being used for different animals and plants and confusion because a single organism has been described differently by different biologists and given two names.

Describing and naming organisms is a highly specialized and difficult part of biology. Now, while it is not appropriate to go deeply into this area, it is necessary to understand the conventions in use in naming and describing.

The first convention concerns the use of the terms 'animal', 'plant', 'organism' and references to named examples of organisms: 'earthworms' for example, or 'fleas' or 'buttercups' or even 'the earthworm', 'the flea' or 'the buttercup'.

Generally such references are to a particular *group* of organisms not to a *single* individual nor to *all* 'earthworms' or *all* 'fleas'. This group is composed of individuals very like each other. No two individual organisms are ever exactly like each other, because of the effects of mutation, recombination and reassortment of the genetic material of partners in sexual reproduction (S100, Unit 19). And in so far as sexual reproduction is the mode of reproduction for the majority of animals, the generalization is extended to all organisms. (Organisms which never or seldom reproduce sexually are difficult to deal with in this argument—see Section 1.1.2 for further discussion.) However, if individuals are able, in spite of their individual differences, to breed, it is assumed that their genetic organization is more rather than less similar. Consequently the basic unit of the biological world—from the present standpoint—is a group of more or less similar individuals which can or do interbreed. This unit is called a *species*—and **species** this unit is the unit described and named by biologists—see Section 1.1.2 for further discussion (and S100[1]). By extension, the name is used for all the individuals making up the group. So a reference to 'the flea' may mean an individual flea or it may be a shorthand way of referring to a species of flea of which one in particular is the one meant. The species name is usually based on a Latin, or Greek or a Latinized word or words. This convention arose because Latin was, when this particular naming system started, a universal scholarly language. The name is usually divided into two parts, a noun-like word plus an adjectival word. The first of the pair is described as the generic name, the second as the specific name. For instance the House Sparrow is *Passer domesticus* L. Note that, again by convention, the generic name is spelt with an initial capital letter but the specific name is usually spelt with an initial lower case letter. Very often the two names are written in italics. The two-word combination or 'binomial' belongs uniquely to the organism to which it is first assigned by the biologist writing the description. The name of the biologist is also recorded—in the example of the House Sparrow, the biologist was Linnaeus, his name being abbreviated to L. The binomial may not be used again ever, for naming purposes, unless an animal name is used for a plant or a plant name used for an animal. But this practice of double use of binomials is not widespread, fortunately. While the full binomial may not be reused, the generic and specific parts of the binomial may be. But in rather different ways: using the same generic name implies that the organism belongs to the same genus, using the same specific name, e.g. *vulgaris* (= common), is merely a convenient 'adjectival' description. For example, primrose and cowslip are respectively *Primula vulgaris* Huds. and *Primula veris* L. The common generic name indicates the close similarity of the two species and their membership of the genus *Primula*. A *genus* may contain a number of species or it may only consist of one species. In a similar fashion, similar genera are grouped to form a family, similar families make up an order, similar orders a class and similar classes either a phylum (for animals) or a Division (for plants). Each one of these categories is referred to as a 'taxon'; each taxon can be ranked with respect to the others in a hierarchy of generality. Organisms with the greatest degree of similarity are arranged in taxa low in the hierarchy—put in the same species or genus for example. Organisms with less similarity may share the same high-order taxon. For example, all animals with the following features (all given in S100, Unit 21):

taxon

> segmented body
> axial notochord
> dorsal, tubular hollow nervous system
> paired gill slits
> bilateral symmetry

are members of the phylum *Chordata*—this phylum includes *all* mammals, birds, reptiles, fishes, amphibia, lampreys, lancelets and tunicates, but excludes *all other organisms*. As in the case of the binominal, the high taxa are named with a Latin, Greek or Latinized word.

In our own case—all men are *Homo sapiens*,
> our family is the Hominidae,
> our order is the Primates,
> our class is Mammalia,
> our phylum is Chordata.

It is often convenient to refer to taxa as being of the same rank or of different ranks—meaning by this that the taxa are at the same level of generality in the hierarchy or of different levels. This discussion is developed further in Unit 2. The practice of naming and describing organisms is subject to an international code of practice. The practice is complex and it is unneccessary to go into detail here, but you should know the function of the generic and specific names. Further than this it is only necessary to understand the hierarchical system of higher order taxa.

1.1.2 Some complications

The simplest complication arises from the fact that some taxa are in practice inconveniently large or appear to be so large as to be meaningless. This is dealt with by dividing phyla into subphyla and classes into subclasses, and so on. The class Mammalia for instance is divided into three subclasses—(ours is the Eutheria).

Subdivision of taxa may become necessary if the size and meaning of the similarities between members of any particular taxon above the species level is considered. It is important to remember that the theoretical criterion for the establishment of a species as an entity is the ability of individuals to breed together. This is a behavioural criterion and can be tested, but in fact it has not been tested for the vast majority of living organisms. For obvious reasons it would be impossible to apply this criterion to extinct organisms and fossils. The test is an ideal. In practice, the identification of organisms, description of species and so on (i.e. the work of the taxonomist) is often carried out on dead organisms in museums. The breeding test cannot be carried out. So in practice other criteria are used—generally depending on the skill of taxonomists in judging closeness of similarity between organisms thought to be members of a species. Similar problems, as far as the breeding test is concerned, are posed by organisms which habitually reproduce by non-sexual means, commonly by vegetative reproduction† in plants, by fission† in the unicells and parthenogenesis† in animals. Once again the refined judgement of the taxonomist is called into play in assigning specimens to species.

criteria for species definition

† *See the glossary, p. 32, for terms marked thus.*

The definition of the term *species* evidently depends very much on an individual taxonomist's viewpoint. The extremes can be represented by those who hold that the species is a genetic unit and those who hold that the species is a phenotypic unit. Both extremes are useful but limited.

Taxonomic systems above the species level may be either 'natural' or 'artificial'. A 'natural classification' is one that reflects its author's beliefs as to the evolutionary history of the organisms concerned. The credibility of such systems depends on the skill and experience of the people who construct them; there can be no objective tests of their validity since it is impossible to be certain of lines of descent except, perhaps, in the recent past. Artificial classifications are constructed purely for convenience in identification and naming; by definition, they carry no implications about interrelationships of taxa above the level of genus or sub-genus.

natural classification

The problem of constructing a natural classification is developed in Section 1.2 and in Unit 2. The criteria for uniting species into a genus or genera into a family are agreed by a consensus of opinion. This being so it is not surprising to find that reshuffling of species into genera and so on occurs from time to time as opinions change.

1.1.3 Conclusion

The description and naming of organisms is both essential and convenient for all biologists. The development of hierarchies of taxa and the classification of organisms is an important part of biological studies but it is a specialized discipline. If you are very interested, you can consult the following technical books:

V. H. Heywood (1967) *Plant Taxonomy*. Studies in Biology No. 5. Edward Arnold. 60 pp.

E. Mayr (1963) *Animal Species and Evolution*. Harvard University Press: Oxford University Press. 814 pp.

R. R. Sokal and P. H. A. Sneath (1963) *Principles of Numerical Taxonomy*. W. H. Freeman. 359 pp.

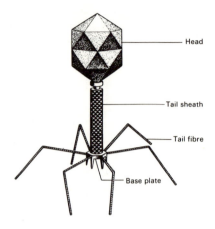

Figure 1 Diagram to show the structure of T_2 virus. (Magnified approx. 10^5 times)

1.2 Survey of Living Organisms

Study Comment

The important parts of this Section, the parts you should learn, are:

1 the characters which distinguish Eucaryota and Procaryota;
2 the five kingdoms into which living organisms are classified.

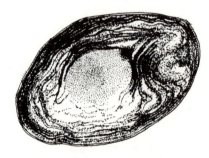

Figure 2 Diagram to show the structure of a photosynthetic bacterium. The lamellae carry the photosynthetic pigment. (Magnified approx. 3.5×10^4 times)

Older classifications divided organisms into two groups, the plant and animal kingdoms (the viruses, bacteria and fungi, as well as the more familiar plants, were classified in the plant kingdom). The animal kingdom has remained virtually unchanged, but the plant kingdom has undergone drastic revision.

Viruses are no longer regarded as plants in modern systems of classification for many reasons; one is that they have a crystalline rather than a cellular structure. Refer back to $S100^2$ to remind yourself about viruses. See Figure 1.

Many authors exclude viruses from the plant kingdom but include *bacteria*. This common practice is open to question because bacteria differ remarkably from all organisms except the *blue-green algae*.

Turn to Appendix 2 (Black) of S100, Unit 18: READ pp. 59 (starting at '1 *Viruses*'), 60 and 62 (as far as '(c) . . . organelles, the choroplasts'). Some authors regard the presence or absence of membrane-bound organelles (nuclei, mitochondria and chloroplasts) as a fundamental difference between groups of organisms. Those cells that lack membranes (Fig. 2) are called 'procaryotic', those that have these membranes are termed 'eucaryotic'. Thus bacteria and blue-green algae fall into a division *Procaryota* while all other organisms (except viruses) fall into another division *Eucaryota* (Table 1).

Procaryota

Modern systems of classification which exclude the viruses, bacteria and blue-green algae from the plant kingdom, frequently also exclude the fungi. *Fungi* lack plastids† (shown in Fig. 3) and in this respect resemble animals more closely than plants.

The fact that animals and fungi both lack plastids does not necessarily imply that the two groups of organisms are closely related. But any attempt to argue a close relationship between fungi and plants must take account of the fact that plastids are absent in one group and present in the other.

In view of the relatively small number of different types of membrane-bound organelles present in eucaryotic cells, the presence or absence of plastids is likely to be a very important difference in evolutionary terms. Because fungi and animals lack plastids, they are unable to utilize light energy to drive synthetic reactions and therefore depend for energy on an external supply of organic compounds. In other words, fungi and animals are heterotrophes (S100, Unit 20). Plants, which are capable of photosynthesis, are autotrophes.

Fungi have traditionally been classified as plants because they have cell walls, but the chemical composition of fungal cell walls is quite different from that of plant cells with plastids. Unlike organisms with plastids, fungi have many nuclei in a single cytoplasmic mass.

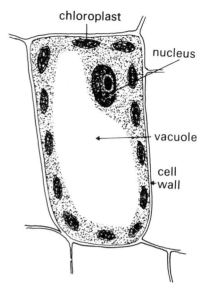

Figure 3 Diagram of a 'typical' plant cell.

TABLE 1

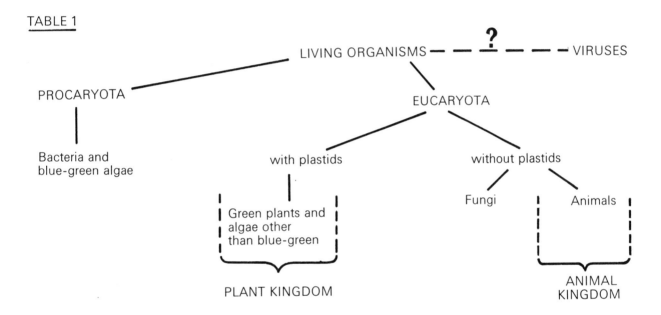

To recapitulate: some systems of classification recognize five kingdoms—the viruses, the bacteria and blue-green algae, the fungi, the plants and the animals—others recognize only plants and animals and still others deal in terms of something in between.

Since all biologists recognize the five distinct groups and differ only on whether the differences between them are sufficient to warrant establishing a distinct kingdom for each, it is reasonable to wonder whether it matters how the diverse groups are classified. If all biologists recognize the five groups, then surely the argument is trivial—or is it? The two-kingdom system of classification originated because people thought in terms of the two most obvious types of organisms—plants and animals. Organisms such as bacteria fairly obviously did not have the characteristics of animals, so they were assigned to the plant group as a matter of convenience. In this way, the Plant Kingdom accumulated all groups of organisms that were obviously not animals. The plant kingdom became the 'not-animal' kingdom. The heterogeneity of the plant kingdom, conceived of in this broad sense, contrasts with the homogeneity of the animal kingdom.

If all that is expected of a classification system is a method of filing organisms, then this heterogeneity of the plant kingdom would not matter. Biologists in general, however, are not very interested in arbitrary systems of classification, even when they can be made to work. They are interested in developing systems which reflect evolutionary relationships between organisms. The closer two

11

organisms or groups of organisms appear in a natural or phylogenetic system of classification, the closer is their implied relationship. This being the case, it does matter whether you settle for two kingdoms or five. If viruses, bacteria and blue-green algae, fungi and plants (in the restricted sense of eucaryotic organisms with plastids) are all included within the plant kingdom, then the implication is that these four diverse groups of organisms resemble each other more closely than they resemble animals. There is no justification for such an implication.

In a comparative physiology course, any number of comparisons are possible. Obviously, some will be more meaningful than others. In general, the simplest and clearest comparisons will be between closely related organisms or homogeneous groups of organisms. This is because it is possible to generalize about closely related organisms or homogeneous groups of organisms. It is much more difficult to generalize about a heterogeneous collection of organisms.

If plants are conceived of negatively, as all living organisms which are 'not animal', it is extraordinarily difficult to make positive generalizations about them. Comparisons between plants (in the broad 'not-animal' sense) and animals are invariably complicated and hedged about with all sorts of exceptions. It is far simpler and more meaningful to compare animals with other equally homogeneous groups. For the purposes of comparing plants and animals in this Course, plants are conceived of in the narrow sense, as *eucaryotic organisms with plastids* (Table 1). The Plant Kingdom is discussed in Unit 2.

To define 'plants' and 'animals' as organisms with plastids and without plastids respectively suggests that there is a clear dividing line between the two kingdoms. But in S100,[3] you met 'protistans'—refer back to these Units if you need to remind yourself about them. They are organisms with their bodies not divided into cells, some with plastids and some without; some can function only as autotrophes, and we would have no hesitation in allocating them to the Plant Kingdom, whereas others can function only as heterotrophes and equally clearly fit well into the Animal Kingdom. But what of the organisms that can function either as autotrophes or as heterotrophes, that are sometimes green and sometimes colourless? What of colourless organisms that are heterotrophes but resemble very closely in detailed structure and life history organisms that have plastids? Must these be separated into different Kingdoms? The compromise we adopted in S100 was to put all single-celled organisms (apart from bacteria and yeasts) into the 'Protista'.

It is traditional for some classes of Protista to be included as 'Algae' and studied by botanists whereas other classes are labelled 'Protozoa' and studied by zoologists. Turn to pp. 62, 64, and 66 of S100, Unit 18: the green algae, yellow-green algae, red algae and diatoms are usually allocated to botanists; the dinoflagellates, euglenoids, amoeboids and ciliates, together with small motile green algae are generally included in zoological texts. Protistans represent a radiation of organisms with bodies not divided into cells; they are diverse in structure and physiology. They do not divide logically into plants and animals and we shall not concern ourselves with them in this Course except very occasionally to illustrate special points.

1.3 The Animal Kingdom (excluding Protozoa)

Study Comment

This is a long Section to study and should not be rushed. You should work carefully and attempt the In-text Questions (*ITQs*) whenever these occur. There are no short cuts in this Section. You must learn the major groups of animals and their characteristics.

Study of the fossil record (refer back to S100, Unit 21 to remind yourself about this) suggests that the earliest many-celled animals (called *Metazoa*) lived in the oceans. Later in this Course we shall discuss the evidence from comparative physiology relating to this view. From marine organisms have evolved animals that live in freshwaters and on the land; the physiological problems associated

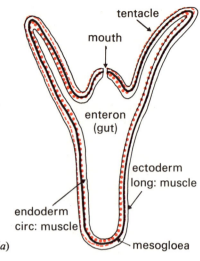

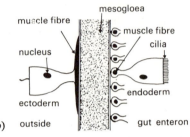

(a)

(b)

Figure 4 Diagrams of a cnidarian polyp: (a) a longitudinal section through the polyp (b) part of the body wall enlarged to show one typical musculo-epithelial cell on each side of the mesogloea.

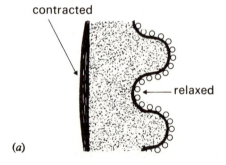

(a)

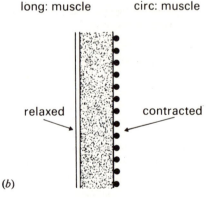

(b)

Figure 5 Diagrams of longitudinal sections through the body wall of cnidarian polyps showing only mesogloea and muscle fibrils; (a) with ectodermal (longitudinal) muscle contracted (b) with endodermal (circular) muscles contracted. (Magnified approx. ×10 times)

with these transitions will be discussed mainly in Unit 10. Habitats of modern animals range from all parts of the oceans to almost all parts of the continents, though animals are rare under conditions of permanent ice or extreme desert.

In size, modern animals range from a few μm to about 30 m long (the Blue Whale). As you might expect from this, they also range in complexity from organisms with comparatively few types of cells arranged in a simple pattern to organisms with many cell types and complex body patterns. In the review that follows, we have chosen to look at a few basic patterns and to study them from the point of view of the movements that the animals can make. We can justify this treatment since movement is a feature that most people would classify as characteristic of animals rather than plants.

In the TV programme of S100, Unit 18, three types of protistan movement were shown: by cilia; by formation of pseudopodia (amoeboid movement); and by changes in body shape (euglenoid movement). Some many-celled animals move by ciliary action, but most of them move by contraction of special cells called *muscles*. (These are discussed more fully in SDT 286, Unit 4.) Contractions of muscle cells cause changes in the shape of organisms and thus movements either of parts of the body or of the whole body.

No muscle can lengthen actively—all muscles lengthen as a result of being stretched by some agency other than themselves; then they can shorten (contract) actively, exerting a force as they do so. In the lives of animals, muscles contract and then relax; the simple fact that a contracted muscle cannot exert further force until stretched passively is a clue to many of the complexities of pattern of animals.

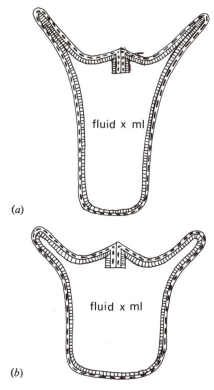

(a)

(b)

Figure 6 *Diagrams of a cnidarian polyp (a) with the circular muscles contracted (b) with the longitudinal muscles contracted.*

1.3.1 Cnidarians (sometimes called 'coelenterates')

In the phylum Cnidaria (refer to *IS* for more detail), one sort of animal, the *polyp*, consists of a muscular body surrounding a fluid-filled cavity, the *enteron*, in which food is digested. In its simplest state, the muscular body is formed of two layers of cells on either side of a layer called the *mesogloea* (Fig. 4 (a)). This has a gelatinous consistency and may contain elastic fibres; there are two membranes, one on either side of it. The cells that lie on either side of the mesogloea are partly muscular and partly either protective or digestive in function (according to whether they are on the outside or the enteron side of the mesogloea respectively); they are called 'musculoepithelial cells'. The muscular part of the cells consist of long fibres that are attached to the membrane of the mesogloea along their lengths (Fig. 4 (b)). In the outer layer of cells, the muscle fibres are arranged 'longitudinally' and in the inner layer of cells, the muscle fibres are arranged 'circularly'. Thus there are two sheets of muscles with the fibres arranged at right angles to each other.

[*IS* Cnidaria A & B]

musculoepithelial cells

> Suppose the fibres of one sheet of muscles contract—what effect will this have on the fibres of the other sheet? See Figure 5.

But such a sheet of corrugated muscle fibres cannot be stretched in any way *unless* it is part of the body of an intact animal *that has its mouth closed*.

If the mesogloea is sufficiently plastic, the other sheet of muscle will buckle and become corrugated.

> **ITQ 1** Examine Figures 4, 5 and 6: explain how the contraction of the sheet of longitudinal muscles results in stretching of the fibres of the circular muscles.
>
> Read the Answer to ITQ 1 (p. 37).

In this very simple polyp, we have noted two features that are characteristic of many animals:

1 There are *two* sets of body muscles that are *antagonistic* to each other (i.e. when one set contracts, the other set is stretched);

antagonistic muscles

13

2   The antagonistic effect results from the fluid in the enteron cavity acting as a *hydro-skeleton*. As long as this fluid remains constant in volume, the system works.

In some Cnidaria, such as the sea anemones, the arrangement of muscles and mesogloea is complicated and there is considerable range of change in shape of the body; but the principle involved is the same as in the simple diagram of the polyp in Figures 4 and 5.

In the Cnidaria there is another type of individual, the *medusa* (jelly-fish); this illustrates a different arrangement of muscles, one in which there are no antagonists to the principal swimming muscles. Look at Figure 7 (a): note that the mesogloea on the side of the enteron away from the mouth is very thick. It has many elastic fibres within it and can be deformed but returns to its original shape because of its elasticity. The swimming muscles are arranged in a circular band near the outer edge of the disc; when these all contract together, they deform the disc, making it more cup-shaped than saucer-shaped (Fig. 7 (b)). As the muscles contract, water is forced out from under the bell and the animal is propelled in the opposite direction.

> **ITQ 2** Suggest what happens when the muscles relax.
>
> Read the Answer to *ITQ* 2 (p. 37).

Of course the jelly-fish has other muscles, round its mouth and in its tentacles; in the latter, there are antagonistic sets of longitudinal and circular muscles just as in the body of polyps (and also in the tentacles of polyps). In the swimming medusa, a single set of muscles work against an elastic 'skeleton': when they are relaxed, the body returns to a definite shape.

In introducing the two types of individual found in the Cnidaria, we have also shown two contrasted habits of life—the *motile* (swimming) jelly-fish and the *sedentary* polyp that lives most of the time attached to some solid object or to the substratum. Some polyps remain attached for the whole of their lives, often as part of a branching colony—these are truly *sessile* animals. Other polyps can move about and a few can swim in a haphazard sort of way by throwing the body into violent bending movements. Polyps were first classified as 'zoophytes' (animal-plants) because of their sedentary habit and flower-like shapes and colours—but most of them are carnivores, capable of catching and swallowing other animals (you will read more about the feeding of polyps in Unit 4).

Polyps and jelly-fish both have mouths and usually a circle of tentacles round these. Because 'head' and 'body' or 'front (anterior) end' and 'back (posterior) end' cannot be distinguished, it is usual to speak of the 'oral' part of a polyp (the part surrounding the mouth), and the 'aboral' part (the part at the opposite end from the mouth). In jelly-fish, the oral surface (under the umbrella) is distinguished from the aboral (or ex-umbrellar) surface. When these animals are viewed from the oral surface, the body is round; cut a section through a polyp at right angles to the oral-aboral axis and it is a circular section.

Look at Figure 8: take a pencil and ruler and draw a diameter—compare the two semi-circles on either side of the line you have drawn.

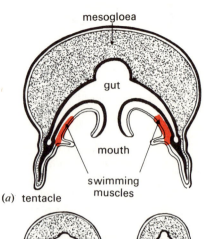

hydro-skeleton

(*a*) tentacle

(*b*) muscles relaxed   muscles contracted

Figure 7   Diagrams of cnidarian medusae: (a) vertical section to show the thick mesogloea and the swimming muscles (in red) (b) vertical sections showing changes in shape during swimming movement.

motile
sedentary

sessile

oral–aboral axis

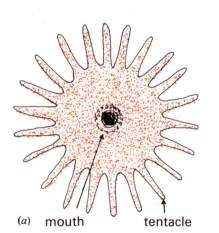

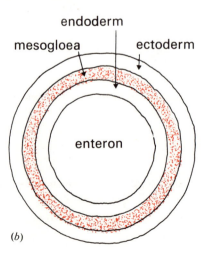

Figure 8   Cnidarian polyp: (a) viewed from above the mouth (b) a horizontal section cut below the oral disc.

Are they identical? Could you cut one out, rotate it and fit it exactly above the other?

Figure 8 shows 'radial symmetry': any diameter divides it into identical halves. Now look at Figure 9 (a): draw diameters through this.

radial symmetry

Are the halves *always* identical? Could you fit each exactly over the other half?

No. No.

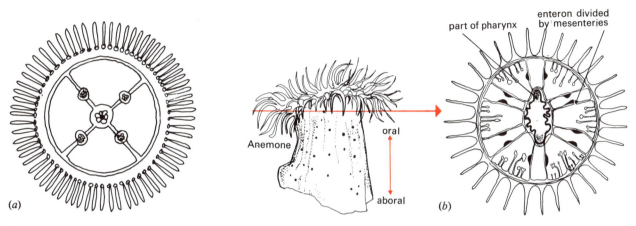

Figure 9 (a) *Medusa viewed from below with mouth in centre of disc.* (b) *Horizontal section of sea anemone cut at the level shown on the drawing of the whole animal.*

On this figure, you can draw two diameters (at right angles to each other) each dividing the circle into identical halves—this is an example of 'tetraradial symmetry'. Jelly-fish conform to this pattern.

Now look at Figure 9 (b): draw diameters through this. How many can you draw that divide the figure into *identical* halves that can be fitted on each other?

Two, but the two pairs of halves are not identical.

This is called 'biradial symmetry'. So here are animals with an oral and aboral end and the other structures arranged in some form of radial pattern round the oral-aboral axis. These animals have a single internal cavity, called the enteron, that may be a complicated shape as a result of the folding of its walls. The body is basically constructed from two layers of cells separated by mesogloea.

The basic characteristics and pattern are so simple that it was assumed in the past that Cnidarians are the most primitive, the first evolved, of many-celled animals and the closest in structure of all living animals to the ancestor of all metazoans.† For various reasons that we have not time to go into here—one of them is that the muscles of Cnidaria are unique in forming sheets attached to mesogloea—we now believe that this group is a branch of the evolutionary 'shrub' quite separate from other many-celled organisms, successful in certain aquatic niches and displaying their own 'peculiar' pattern of organization. So what are other many-celled organisms like? Their muscles are elongated cells attached by their ends to other structures as in Figure 10—not attached along their lengths to a membrane as are Cnidarian muscles (Fig. 4). The difference means that the muscle cells of most animals are able to form the solid structures we usually call 'muscles' in contrast to the flat sheets of muscle fibres of Cnidaria. But physiologically the muscles behave in a similar way, they contract actively and can be stretched passively when they are relaxed. How are they relaxed? How are they arranged in these animals and how do the animals live?

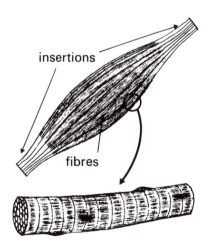

Figure 10 *Diagram to show the structure of typical muscles of metazoan animals other than cnidarian.*

1.3.2 Worms

Most flatworms (*Turbellaria*, phylum *Platyhelminthes*) are small, active, carnivorous animals (refer to *IS* for more detail). Many are ciliated over their
bodies and glide along over the substratum by ciliary action; they alter their
shapes by muscular contractions. Others 'crawl' by muscular activity as a
result of a series of ripple-like waves passing along the lower surface, the part of
the body in contact with the substratum. Look at Figure 11: this flatworm has
eyes near one end of its body and usually moves with this end leading. So we can
describe the animal as having an anterior end and a posterior end—but is the
anterior end equivalent to a 'head'?

This depends on how you define 'head'; but notice the position of the mouth (*not*
at the anterior end but about one-third of the body length behind the leading
edge). The flatworm always crawls with the same surface in contact with the
ground—so we can distinguish between an upper 'dorsal' surface and lower
'ventral' surface. Looking at the pictures of the animal, consider whether it has
any of the forms of radial symmetry shown in the diagrams of cnidarians in
Figures 8 and 9. The flatworm can be divided down the midline into two halves
that are mirror-images of each other. This type of symmetry is called 'bilateral'—
notice as you read on through this text that the majority of metazoan animals
display this type of symmetry and can be said to have anterior and posterior
ends, and dorsal and ventral surfaces.

[*IS* Platyhelminthes A & B]

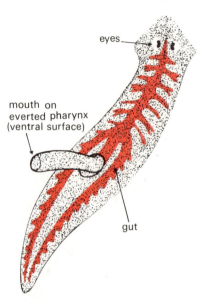

Figure 11 *Diagram of flatworm (pla-
thyhelminth) showing the gut with the
pharynx everted through the ventral
mouth. (Magnified approx. ×10 times)*

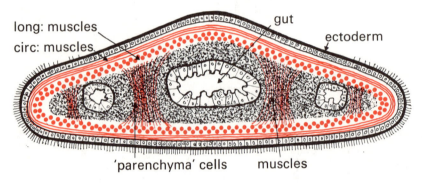

Figure 12 *Diagram of a tranverse section through a flatworm with the principal body
muscles shown in red. (Approx. × 10)*

Figure 12 shows a section cut across the body of a flatworm. Notice the gut
cavity and some tubes forming parts of the reproductive system and excretory
system. The rest of the inside of the body is packed with cells. Some of these are
muscle cells; these are arranged longitudinally and circularly and some run
diagonally, others dorso-ventrally—the muscular system is complex in the larger
flatworms. Between these muscle cells and special tissues such as those of the
digestive and nervous systems there are undifferentiated cells, sometimes with
large fluid-filled vacuoles in them and sometimes with fluid-filled small spaces
between them. These cells and the fluid in or between them function as a hydro-
skeleton, giving the body a fairly constant volume. The complicated muscles
could function as sets of antagonists but probably, when most flatworms
crawl, the system is more like that of a jelly-fish: the longitudinal muscles near
the ventral surface contract in one area, slightly distorting the body; when they
relax, they are stretched as the body regains its former shape (Fig. 13). The
longitudinal muscles contract in sequence along the body and, as this wave
passes along, the animal moves forwards with the relaxed muscle in contact with
the ground and the contracted areas of muscle lifted above it.

Flatworms earn their common name because the body is very thin in the dorso-
ventral plane. These animals lack a blood-vascular system, but have a gut
whose branching pattern is more complex the larger the species. There is no anus
and undigested food is voided via the mouth. Flatworms are quite common
animals in some shallow marine habitats and in some fresh waters. All are
active carnivores but they have relatives, the flukes and tapeworms, that are
parasites for most or the whole of their lives.

parenchymatous hydro-skeleton

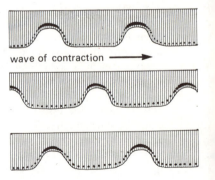

Figure 13 *Diagram to show how waves
of muscular contraction can pass along
the ventral surface of a flatworm.*

16

Figures 14 and 15 show two other sorts of worms: one is the familiar, earthworm (phylum *Annelida*) and the other a roundworm (a nematode, phylum *Aschelminthes*)—let us look at the latter first. The roundworm has a mouth at one end and sensory bristles and pits near it; a few roundworms have a pair of eyes. The anus is near the posterior end. Look at the whole worm and the transverse section through it—check that the type of symmetry is bilateral not radial. The gut forms a tube down the middle and the gonads also are tubular. The muscles are arranged in four blocks under the outer layer of the body—all the muscle fibres are orientated longitudinally. There is a fluid-filled space between the gut wall and the muscle cells—this is called the *pseudocoel*.

[*IS* Annelida A & B]
[*IS* Aschelminthes A & B]

ITQ 3   Suggest how contracted muscles are stretched in this worm.

Read the Answer to *ITQ* 3 (p. 37).

Nematode worms move by sideways bends that are not very efficient except in viscous fluids through which the worms can glide; many of them are parasites but many are free-living, especially in the soil or in bottom deposits of the sea and fresh waters.

Look now at the transverse section of the earthworm (Fig. 15):

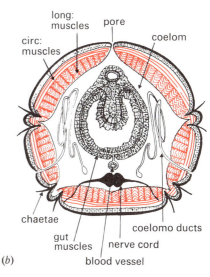

(a)

(b)

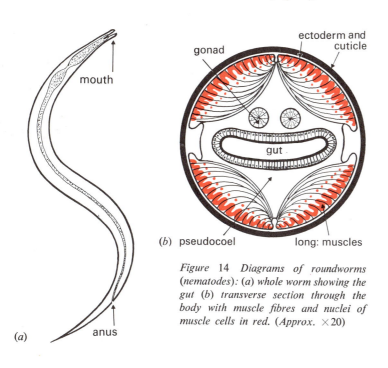

Figure 14 *Diagrams of roundworms (nematodes):* (a) *whole worm showing the gut* (b) *transverse section through the body with muscle fibres and nuclei of muscle cells in red.* (*Approx.* ×20)

Figure 15 *Diagrams of earthworms (annelids):* (a) *whole worm* (b) *transverse section through body with principal muscles in red.* (*Approx.* ×7)

ITQ 4   How many sets of muscles are there? Is the gut the only cavity shown in section inside the worm?

Read the Answer to *ITQ* 4 (p. 37).

The difference between the earthworm's coelom and the pseudocoel of a roundworm is that the coelom is a cavity between two layers of muscle (the gut muscle and muscle of the body wall) each covered by a layer of epithelium cells. The gut can undergo peristalsis independently of movements of the body wall. In the roundworm, the pseudocoel lies between the gut wall (which is not muscular) and the body muscles; changes in shape of the gut result from body movements and are not independent. The fluid in the pseudocoel bathes the outside of the gut and the inner borders of the muscle cells directly; there is no epithelium. These may seem small differences but they are tied up with differences in early development that appear to zoologists to be very important when discussing relationships between groups of animals.

Figure 16 shows a longitudinal section along part of the body of an earthworm—note that the coelomic cavities are separated from those anterior and posterior

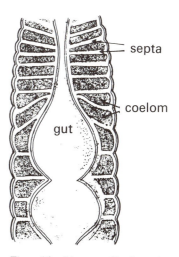

Figure 16 *Diagram of horizontal section through the body of an earthworm near its anterior end to show coelom divided into segments by septa.* (*Approx.* ×4)

17

to them by sheets of tissue called *septa*. The body muscles, also, are arranged in a series of blocks along the body. This breaking-up of the coelom and muscles into blocks is called *segmentation* (you met the term in S100, Unit 21 applied to the muscles of vertebrates).

ITQ 5 Consider the muscles in one block (segment):
(a) Suppose that the longitudinal muscles contract and fluid in the coelom remains at constant volume, what will happen?
(b) Now suppose that the circular muscles contract, what will happen? Look at Figure 17.

Read the Answers to *ITQ* 5 (p. 37).

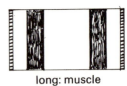

(a) long: muscle
contracted

(b) circ: muscle
contracted

Figure 17 *Diagrams to show shapes of earthworm segments* (a) *with the longitudinal muscles contracted* (b) *with the circular muscles contracted.*

Look back to the Section on Cnidarians and compare this system of cavities and muscles with that of the polyp (Fig. 6): here is the same sort of antagonism between sets of muscles arranged at right angles to each other—even though the details of the musculature are very different. The coelom of the earthworm functions as a hydro-skeleton exactly as the enteron cavity does in the polyp.

In a live earthworm, you can observe that contractions of the longitudinal and circular muscles pass alternately along the body which is fixed to the ground where the longitudinal muscles are fully contracted and moves forward between these points. The segmental arrangement of muscles and coelomic cavities allows rather precise control of locomotory activity, each segment working as a unit within the body.

In studying an annelid, such as the earthworm, we have met several features which are characteristic of the majority of many-celled animals: bilateral symmetry; body with anterior end bearing the mouth and, often, principal sense organs; coelom present; and main body muscles segmented. In annelids, the coelom acts as a hydro-skeleton but, as you will see later, other phyla have more orthodox types of skeletons.

Look back to Figure 15, to the cross-section of the earthworm. We have referred to two sets of cavities: the gut and the coelom. The gut opens to the outside at the front end through the mouth and at the back end through the anus. The segmented coelomic cavities are also connected to the outside: by dorsal pores in each segment and through tubes (called *coelomoducts*) which convey certain products of the coelomic walls to the outside. Typically, sperm and ova are produced in coelomic cavities and reach the outside through coelomic ducts; these ducts often have an excretory function as well.

There is another series of tubes in Figure 15—the blood vessels. What relation have the cavities of these tubes to the other cavities we have just mentioned? The blood vessels of the earthworm are entirely separate from the gut cavity and from the coelomic cavity; they represent a space which appears early in development and eventually becomes the cavity of the blood vessels. In the worm these form a 'closed system'. (You will read about this and other types of blood system in Unit 5.)

Lest you should gain the impression that all annelid worms resemble the earthworm, look at *IS* Annelida C and E (polychaete worms and leeches). Compare *Nereis* with the earthworm: look at its front end and at the shape of the body segments. The earthworm has a smooth body but in *Nereis* there are a series of 'appendages', one pair per segment; on its front end there are eyes and modified appendages. Look at Figure 18 (TS* *Nereis*) and compare it with Figure 15 (TS earthworm). *Nereis* moves by lateral undulation of the body—a type of movement like the side-to-side bending of a fish body.

Which are the muscles involved?

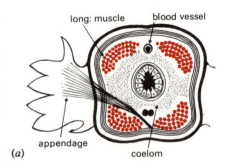

(a) long: muscle   blood vessel

appendage   coelom

coelom   R
L

L contracted

L   coelom
R

R contracted

(b) L and R are longtitudinal muscles of two sides of body

Figure 18 *Diagrams of the ragworm Nereis* (annelid): (a) *transverse section through body with longitudinal muscles in red.* (Approx. ×7) (b) *shapes of segments with left and right longitudinal muscles contracted.*

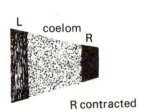

The longitudinal muscles.

Here the antagonism is between the longitudinal muscles of the right and left sides of the body. Assuming that the *volume* of each segment remains constant, then contraction of the two blocks of longitudinal muscles on the left side must

* *Transverse Section.*

18

stretch the two blocks on the right side and vice versa (Fig. 18 (b)). Contractions pass along the worm alternately on the two sides of the body—an actively swimming worm may be thrown into five or more waves simultaneously. In swimming, the waves pass forwards along the worm and the appendages are held at right angles to the body. When crawling, the worm may move its appendages by contractions of the muscles inserted in them.

Nereis crawls over the bottom of the sea or under stones or swims for short distances, it may also burrow by crawling into mud. Its habits differ from those of the earthworm; it is carnivorous and a more active animal than the worm. The fanworm *Sabella* lives a very different life from either *Nereis* or the earthworm. It moves up and down its tube but never (normally) leaves it. As you will read in Unit 4, it feeds by extracting small particles from the water, the mechanism depending on the rigid tentacles being held up into the water out of the tube. Such a succulent morsel would not survive for long were it not able to withdraw rapidly into the tube. Look at Figure 19: compare the musculature of *Sabella* with that of *Nereis*. The extensive coelom provides a hydro-skeleton; when all the longitudinal muscles contract the whole worm withdraws rapidly into its tube; it moves up the tube again slowly by a (peristaltic) movement exerting friction against the tube with its chaetae. The tentacles are rigid as a result of a central core of cells which are very turgid and act as a skeleton—compare these with the mesogloea of the Cnidaria.

The leech (Fig. 20) looks more like an earthworm than a polychaete worm because its body is smooth, but it has suckers at each end. It swims by undulations of the body or moves by 'looping' as shown in the diagram. Look at Figure 20: try to answer the following questions.

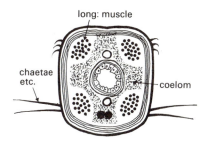

Figure 19 *Diagram of a transverse section through the body of the fanworm* Sabella (*annelid*). (*Approx.* ×10)

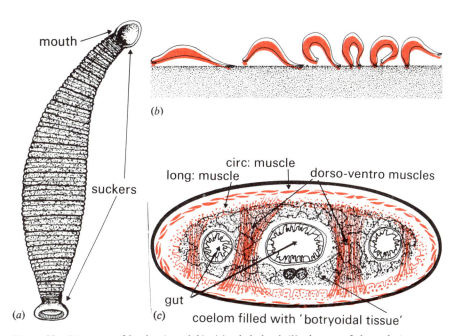

Figure 20 *Diagrams of leeches* (*annelids*)*: (a) whole leech (b) changes of shape during 'looping' (c) transverse section through body with principal muscles in red. (Approx.* × 10)

ITQ 6 (a) What sort of changes in body shape would you expect: peristalsis, horizontal bending or vertical bending?

(b) What are the muscles involved in this?

(c) Which of these muscles are the antagonists?

(d) What part of the body could be acting as the 'skeleton', allowing one of the sets of muscles to stretch the other?

Read the Answer to *ITQ* 6 (p. 37).

In leeches, the coelom is almost filled with cells that are usually turgid with water. It is this 'botryoidal' tissue that acts, among other functions, as a hydro-skeleton. Compare Figures 20 and 12—notice the similarities between flatworms and leeches in general layout, but also that there are differences, especially in muscularity and the presence of blood vessels in leeches.

This similarity between flatworms and leeches is an example of convergence.

All the annelid worms have segmented bodies and there are three systems of spaces within these bodies: gut, coelom and blood system. Compare these and the other worms you have met so far by filling in Table 2.

Table 2

	Flatworm	Roundworm	Earthworm	Nereis	Sabella	Leech
Is a gut present?						
Has the gut muscles in its walls?						
Is a pseudocoel present?						
Is a coelom present? (a) filled with fluid (b) filled with cells						
Are circular body muscles present?						
Are longitudinal muscles present?						
Which muscles are involved in main movement of body?						
Are blood vessels present?						

Compare your version with that given on p. 38.

There are a number of small phyla whose members live in a similar way to the fanworm *Sabella* and show similarities to it in external form and, superficially, in internal anatomy. Here we shall discuss one of these only—*Cephalodiscus*, a member of the phylum *Hemichordata* (some of which are called 'acorn-worms'). This small animal (Fig. 21) feeds with its tentacles extended; it is able to withdraw very rapidly when disturbed and then extends slowly. The tentacles are kept rigid by a hydro-skeleton that is a part of the collar coelom and separate from the main body coelom; functionally this is equivalent to the tentacular skeleton of *Sabella*. There is a third type of coelom cavity (in the shield) in addition to the cavities of the tentacles and the extensive coelom round the gut. Technically, this animal is *unsegmented*; it has a limited number of coelomic cavities in contrast with the many (segmented) coelomic pouches of *Sabella*. *Cephalodiscus* also has a blood-vascular system.

[*IS* Hemichordata A & B]

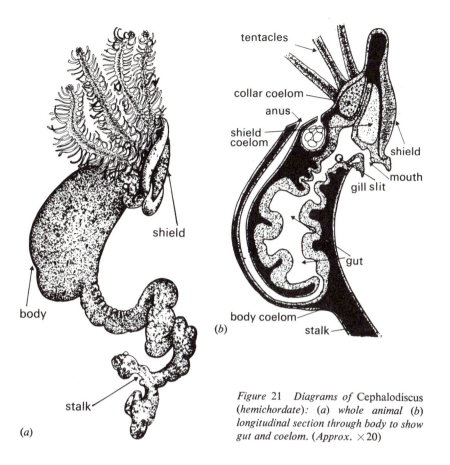

Figure 21 Diagrams of Cephalodiscus (hemichordate): (a) whole animal (b) longitudinal section through body to show gut and coelom. (Approx. ×20)

1.3.3 Molluscs

Figure 22 (a) is a diagram of a snail. Notice the division of the body into a 'head-foot' (with mouth and tentacles at its anterior end) and 'visceral mass' (coiled gut and 'liver' lying in spiral coils of the shell). When you watch a snail moving over a glass plate you will notice ripples passing along the sole of the foot (these are shown very well by water snails crawling on the wall of an aquarium); the ripples are contractions of the longitudinal muscles.

[IS Mollusca A & B]

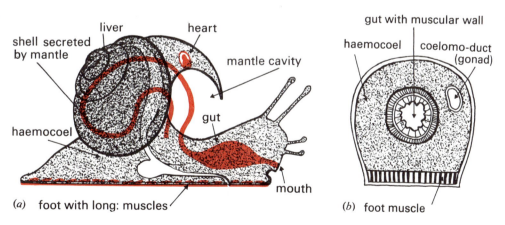

Figure 22 Diagrams of a snail (mollusc): (a) longitudinal section through body to show gut (red), gonad (white) and haemocoel (light stipple) (b) transverse section across foot to show cavities and longitudinal muscles.

This ought to have reminded you of flatworms crawling over surfaces.

Look at the TS of the snail in Figure 22 (b). Suggest how the longitudinal muscles of the foot are stretched.

The fluid in the *haemocoel* acts as a hydro-skeleton with the volume of the head-foot remaining constant. This fluid acts in a similar way to the parenchyma cells of the flatworm.

haemocoel

What is the haemocoel? The word means 'blood cavity' and explains the origin of this space admirably: the haemocoel is equivalent to the expanded cavity of blood vessels and the fluid within it is blood; it is connected with blood vessels.

Look again at Figures 22 (a) and (b) to identify further cavities: note the gut and tubular parts of the gonads (these are equivalent to coelomoducts of annelids). Also shown is the heart (whose cavity is part of the blood system, equivalent to haemocoel) and a space round it, the pericardium (part of the coelom). Thus in this mollusc there are present all the types of space we noted in annelid worms such as the earthworm: gut, coelom, blood system (equals haemocoel). But the snail and the earthworm differ in the relative proportions of coelom and haemocoel. Both cavities carry out several functions in both animals. In Table 3 assign the groups of functions A, B and C to columns 1 and 2, using each function twice, once for the snail and once for the earthworm.

Table 3

	Column 1 Coelomic fluid (including cavity of gonad)	Column 2 Fluid in haemocoel or blood vessels
Earthworm		
Snail		

Insert in the appropriate column for each animal:

A for functions of blood as a transport system for products of digestion, respiration, etc.;

B for functions in relation to formation and discharge of products of gonads (sperm, ova);

C for functions in relation to locomotion as a hydro-skeleton.

Compare your version with that on p. 38. From this table it appears that there are certain functions which the two cavities have in both animals but that the function as a skeleton is carried out by the larger of the two cavities—the coelom of the earthworm and the haemocoel of the snail.

Note that the snail is *not segmented*—neither the main body muscles nor the coelomic cavities are divided into separate blocks. The early development of some molluscs and some annelids is very similar but there comes a stage at which segmented blocks, each surrounding a coelomic cavity, appear in the annelid, whereas the mollusc continues to develop without segmentation.

The phylum Mollusca includes a great diversity of form but all can be derived from a basic pattern (shown in *IS* phylum *Mollusca* B). Those features which are characteristic of molluscs (found in most members of the phylum) are: body with coelom present but not extensive (represented by cavities of pericardium, gonads and kidneys): body not segmented; body consisting of head-foot and visceral mass; mantle cavity present with a pair of gills within it. The visceral mass is often surrounded by a shell secreted by the outer layer of the mantle. This shell may be single (as in snails) or consist of two valves (as in cockles and mussels) or be of many parts; its function generally seems to be protective but it can act as a firm origin for muscles such as those that pull the body of a snail into its shell or shut the valves of a mussel. Thus snails have a hydro-skeleton (the haemocoel) and an external shell.

The molluscs that move fastest are the squids, members of the class Cephalopoda. Look at Figure 23. The head-foot is represented by tentacles, head and funnel. The squid can move slowly by means of undulation passing along the lateral fins. It moves fast by a type of jet propulsion expelling water from the mantle

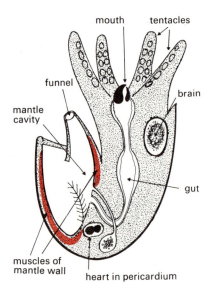

Figure 23 *Diagrams of a squid (mollusc) cut in longitudinal section to show position of mantle cavity and muscles (red) in its walls. (Approx. $\times \frac{1}{3}$). The animal usually lives with the mouth anterior and mantle ventral (i.e. rotated through 90° compared with this diagram).*

[*IS* Mollusca B]

jet propulsion

22

cavity through the funnel, which can be pointed forwards or backwards. The sudden contraction of muscles in the mantle wall creates the jet of water; the muscles are stretched again by the elasticity of the thick walls of the cavity.

Would you class squids as having a very similar swimming mechanism, a not very similar or a very different mechanism from that of Cnidarian medusae?

Very similar.

In the Cephalopods the haemocoel is almost obliterated and the blood flows in vessels. The coelom is small, as in snails.

1.3.4   Echinoderms

The phylum Echinodermata is possibly unfamiliar to you; all its members live in the sea, some on the sea-shore, some on the sea bottom down to very great depths. These animals have many peculiar characteristics but they share with the molluscs the features of having coelomic cavities and being unsegmented. They also have a skeleton, typically of many plates which may be loosely (starfish) or closely (sea-urchin) fitted together; this skeleton is *internal*, lying under the outer layer of the body. Look at Figures 24 and 25 which show these features. Note that several different coelomic cavities are present but there is no definite blood system (or haemocoel).

[*IS* Echinodermata A & B]

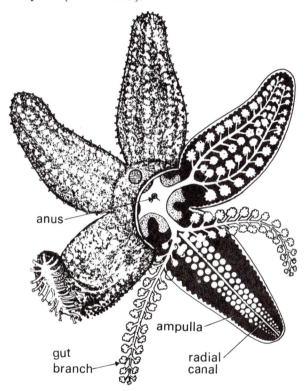

anus
ampulla
gut branch
radial canal

Figure 24   *Diagram of a starfish (echinoderm) seen from above with two arms dissected to reveal parts of gut, ampullae of tube feet and radial canal. (Approx.* × ½)

Look at the starfish in Figure 24. What is its symmetry? Locate the anus; the mouth is below it. We can define here an oral surface, applied to the substratum and an aboral surface on the opposite side.

In the phylum Cnidaria, we met examples of radial, tetraradial and biradial symmetry. Given that most starfish have five arms and many other echinoderms can be derived from a five-rayed state, suggest a term to describe their symmetry. Rendering 'five-rayed form' in Greek gives the term *pentamerous* (used generally instead of 'penta-radial' which is an equally good descriptive term).

pentamerous symmetry

Observations on starfish have revealed that they may move with any of their five arms leading or with the space between two arms leading and may change over to any of the other arms or spaces between them after about fifteen minutes. They move as a result of co-ordinated activity of organs called *tube feet*; these are found only in echinoderms and their presence is a diagnostic feature of this group. Consider now how they work. Look at Figure 25.. Identify the cylinder of longitudinal muscle of the tube foot by which the disc at its end can be lifted up or tilted and pointed in various directions.

tube feet

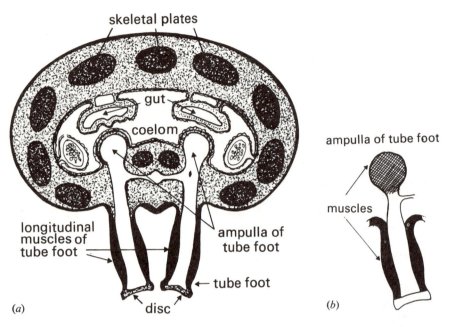

Figure 25 *Diagrams of starfish (echinoderm): (a) transverse section through arm to show cavities of gut and coelom (b) coelomic cavity and muscles of tube foot shown enlarged.*

ITQ 7 Suggest how these muscles might be stretched when relaxed.

Read the Answer to *ITQ* 7, p. 37.

The body remains fairly rigid and is moved along by the 'stepping' of the tube feet of the whole body; these swing forwards and backwards along the direction of movement. This direction may be along the length of an arm or across it at right angles or obliquely. The rate at which starfish move is not very rapid, but they are carnivores, searching out and eating mostly bi-valve molluscs.

Are echinoderms segmented animals? You may think that the arrangement of ampullae of the tube feet is sufficiently similar to that of the coelomic cavities of the earthworm to make the answer 'yes'. But these cavities are only a small part of the coelomic system of the starfish. Notice the very large perivisceral coelom which is the main space inside the body; this is not divided up into linear compartments. The early development of echinoderms is different in various ways from that of annelids or molluscs; there is no breaking-up of coelom or muscles into anything resembling the segmented coelomic cavities of annelids. We therefore class the echinoderms, like the molluscs, as coelomate but not segmented animals. The presence of several different sorts of coelomic cavities is a resemblance between echinoderms and the small phyla represented by *Cephalodiscus*, although the habits of echinoderms such as the starfish are very different from those of the others.

1.3.5 Arthropods and vertebrates

Two major phyla remain for this survey; both were considered in S100, Unit 21, the vertebrates in some detail and the arthropods (with the molluscs) very briefly. Arthropods and vertebrates both display *segmentation*, though this is more obvious from the outside in the former than in the latter. Members of both phyla are typically *bilaterally symmetrical* and clearly have an anterior end, often with a distinct head, bearing sense organs; typically they are active, motile animals. In both phyla, the members have a *gut*, with a mouth on the head

[*IS* Arthropoda A & B]

24

and an anus which is at the hind end of arthropods but in front of the tail of vertebrates. *Coelomic* cavities are present in both as also is a *blood system*. Thus for both phyla we can list the following characteristics: they are bilaterally symmetrical, segmented coelomate animals; but if we look further at the structural patterns of their bodies, we find many striking differences. Compare Figures 26 and 27 and answer the following questions:

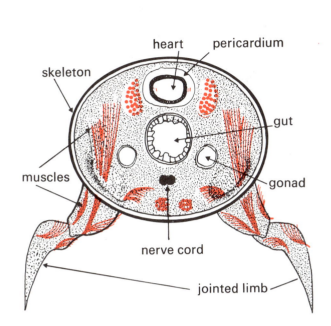

Figure 26   Diagrammatic transverse section through an arthropod to show exoskeleton, body muscles (in red) and cavities in the body. (Approx. × 3)

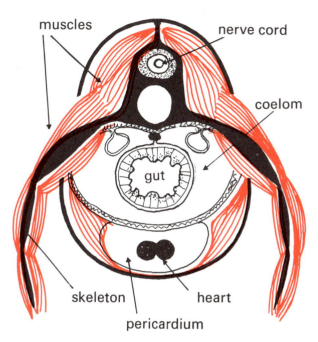

Figure 27   Diagrammatic transverse section through a vertebrate to show endoskeleton, body muscles (in red) and cavities in the body.

ITQ 8   Locate the *skeleton* of each—are they in similar positions in relation to the muscles or in different positions?

ITQ 9   Identify the main *body cavity* (the largest in the diagram)—is it the same in both?

ITQ 10   Locate the *nerve cord* of each—are these in the same position in relation to the gut? Are the nerve cords of the two similar in these sections?

Read the Answers to *ITQ*s 8–10 (p. 37).

Members of the phylum Arthropoda, with their external skeleton, can generally be recognized by the jointed structure of their limbs and the joints between their body segments; there are, of course, exceptions such as some of the maggot-like larvae of insects which lack legs and resemble worms in general appearance. In S100, Unit 21, you read about some of the variety of arthropods—turn back now to Section 21.5 and look at Figure 29 and at filmstrips 21a and b to remind yourself about these. Note that the crustaceans are mostly aquatic animals, some swimming and some crawling on the bottom of the sea or freshwaters; the insects, centipedes, millipedes and spiders are mostly terrestrial animals, able to run on land; only the insects develop wings, as adults, and can fly. The muscles which bring about all these movements are attached to the skeleton, usually with the origin on one side of a joint and the insertion on another side of it (Fig. 28).

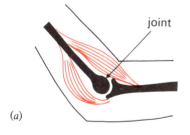

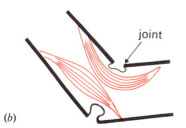

Figure 28   Diagrams comparing (a) endoskeleton with (b) exoskeleton, showing joints and muscles (in red).

ITQ 11   Do you expect a hydro-skeleton to be essential for successful movement in such an animal?

Read the Answer to *ITQ* 11 (p. 37).

Fluid pressure in the haemocoel is important in some arthropods for limb and body movements, but it is not essential with the arthropod type of organization.

25

For successful swimming or walking or running, there must be co-ordination between the various limbs; the segmental structure of the body probably contributes to the efficiency of this since waves of activity can pass in a controlled way along the body. If you have the opportunity, watch a live millipede crawling and consider some of the problems involved in co-ordination of the eighty or more pairs of legs.

To remind yourself about the vertebrates, look at S100, Unit 21, Sections 21.2, 21.3 and 21.4. Part of the TV programme of Unit 21 was about swimming and other types of locomotion of vertebrates. You can read more about them in *M & V*, Volume 1.

> **ITQ 12** How essential is fluid pressure within the coelom for successful swimming of fishes or running of mammals?
>
> Read the Answer to *ITQ* 12 (p. 37).

Although their skeletons are very different in position, in structure and in composition, from the point of view of movement and arrangement of loco-motory muscles, there are many similarities between vertebrates and arthropods. In both groups, antagonistic sets of muscles operate in relation to joints between parts of the skeleton; it is the strength of this skeleton in relation to stresses put on it by the weight of the body and by tension in contracting muscles that is vital for successful movement and survival. Although body cavities are present, the fluid pressure in these is not generally important in locomotion. These two groups include all terrestrial animals that can run.

1.4 Phylogeny—the Interrelationships between Phyla of Metazoan Animals

Study Comment

Comparisons between organisms are likely to be meaningful only if the degree by which the organisms are phylogenetically related is taken into account. This Section should be studied to obtain an evolutionary 'overview' of the Animal Kingdom rather than to learn details of phylogeny.

Section 1.3 reviewed many phyla of animals, looking at them from the point of view of how they move (or how they live) and illustrating some of the great diversity of form among them. We suggest earlier that all these groups of animals are probably more closely related to each other than any of them are to members of the Plant Kingdom or to bacteria or to viruses. Can we make closer groupings among the phyla of animals and perhaps reduce the apparent chaos by construction of 'super-phyla'? To try to do this brings out some of our **super-phyla** ideas of the directions in which evolution has proceeded and the kind of criteria by which we judge relationships between animals. All our conclusions must be speculative, since we shall never be able to check, by observation, what actually happened in the distant past when the first many-celled animals appeared on Earth—most probably in the ocean.

Bearing in mind the four generalizations about evolution that we listed in the Introduction to S100, Unit 21 (21.0) we should expect the common ancestor (if such existed) of all metazoan animals to be a small, simple animal; from this, the phyla living today will have diverged by specialization, adaptive radiation and further specialization. In the past, it has been fashionable to arrange phyla in a hierarchy depending on the apparent complexity of their internal structure. Fill in the blanks in Table 4 and deduce from it a possible hierarchical system based on body cavities and muscular arrangements. Then check your Table against that of p. 39.

Table 4

Phylum or Class	CNIDARIA	TURBELLARIA	NEMATODA	'ANNELIDA	MOLLUSCA	HEMICHORDATA	ECHINODERMATA	ARTHROPODA	VERTEBRATA
Animal	Polyp	Flatworm	Roundworm	Earthworm	Snail	*Cephalodiscus*	Starfish	Insect	Mammal
Gut cavity (or enteron)									
Pseudocoel									
Coelom									
Blood system (or haemocoel)									
Segmented muscles									

Insert in each space: either 0   if this feature is not present;
                              +   if this feature is present;
                             + +   if this cavity is the principal body cavity.

27

In your hierarchical system, you should have started with the gut (enteron) as apparently the most primitive body cavity—the only one in Cnidaria and flatworms, it is present in all the others but not as the principal body cavity. The pseudocoel appears to be peculiar to roundworms, suggesting that this group is perhaps a specialized offshoot of the evolutionary 'tree' or 'shrub', without close affinities to the others. The other two types of cavity—coelom and haemocoel (blood system)—are found in almost all the other phyla, but echinoderms lack a blood system, suggesting that these perhaps occupy the next hierarchical level. Echinoderms also lack segmented muscles—perhaps that is a more primitive state than to have segmented muscles? This suggestion is supported by the absence of segmented muscles in Cnidaria, flatworms and roundworms, groups already assigned to lower levels of complexity. So our hierarchical system now looks like this:

grades of structural complexity

Level of complexity			Phylum (or animals)
1		Enteron cavity only	Cnidarians; flatworms
2A	1	plus pseudocoel	roundworms
2B	1	plus coelom	echinoderms
3	2B	plus blood system as subsidiary set of cavities	*Cephalodiscus*
4	3	plus segmented muscles	earthworms; vertebrates
5≡	4	but haemocoel is principal body cavity and coelom is reduced	arthropods

Now we are faced with a problem—where should we place the molluscs? They cannot go into 5 because they lack segmented muscles but they do not fit in 3 because their principal body cavity is a haemocoel, the coelom being small. On this system of levels the molluscs are misfits, they have attained level 5 without at the same time attaining level 4. Like the roundworms (level 2A) they appear peculiar.

Perhaps if we try a different system of hierarchies, we can fit all the phyla into one scheme of progression? Try, if you wish, but you are not likely to be successful. Is this really surprising? The idea of expecting to be able to arrange the phyla in an orderly series of advancing complexity assumes that each major 'improvement' in body pattern evolved only once so that all animals with that feature must have shared as common ancestor the first animals to have that feature. Was a coelomic cavity evolved only once? Were segmented muscles evolved only once? Was a haemocoel evolved only once?

Fifty years ago it was generally believed that the coelom and segmentation represented such major advances in structure that they could only have evolved once; some modern authors still assume implicitly or explicitly that these features must have evolved in exactly the same way (by a similar series of steps) although they may have appeared independently in two or more evolutionary lines. But these assumptions are not necessary. If a feature confers an advantage on the animal possessing it such that this animal has a greater chance of survival, then whenever that feature happens to appear (by mutation), there is a probability that the 'new' type will survive, reproduce and flourish. If the possession of a coelom confers an advantage that is not available to a non-coelomate animal, then any group in which a cavity appears between the muscles of the gut and the muscles of the body wall has a better chance of survival than its non-coelomate ancestors. In fact this cavity, besides being potentially a hydro-skeleton, allows movements of the gut to happen independently of movements of the main body muscles—probably this confers a very distinct advantage. The presence of a haemocoel also allows gut movements but the pseudocoel does not, the roundworm gut having no muscles in its walls. Similarly, if segmented muscles confer an advantage—better co-ordination of locomotory movements, perhaps—then any stock in which such segmentation appeared (by chance) would have a distinct probability of survival. The evolution of segmentation in unrelated groups of animals would thus be an example of convergence.

convergent evolution of major structural features

You can see in S100, Unit 17 that the control of protein synthesis through DNA via different sorts of RNA is a common feature of all organisms and that the genetic code seems to be a universal code. Considerable biochemical similarities underlie the great structural diversity of living organisms. Eucaryote cells have a limited repertoire of organelles; among animals, there is a limited variety of types of cells. During development from the single fertilized egg cell to a complex adult animal, only a limited selection of cell differentiation and cell movements are possible. Thus it is quite possible that coelomic cavities might have evolved independently several times in several different evolutionary lines, all presenting very similar features and conferring very similar advantages.

If the possibility is accepted that similar major features of body organization may have evolved independently in different lines of evolution, how can we judge closeness of relationship and so construct 'phylogenetic trees'? In practice, as many features as possible of the phyla must be considered; then a balance sheet is drawn up from which to assess whether the animals are so alike that it is highly improbable that they are not related, or whether they appear to be similar because they lead similar lives but have sufficient differences to make it unlikely that they are related. This involves access to a considerable amount of detailed information, but it is a fascinating line of study. Embryology, comparative anatomy, comparative physiology and comparative biochemistry are all relevant.

criteria for plylogeny

So can we group into 'super-phyla' the various phyla discussed in Section 1.3? Yes, but the scheme described here is still controversial: if you read many books, you may find as many different groupings of the phyla—and the fashion changes from time to time!

First, it is generally accepted that the *Cnidaria* should be separated from the rest. Probably they represent a separate radiation of metazoan animals characterized by their peculiar muscle cells attached to mesogloea (and also by stinging cells, produced in no other phylum). They illustrate many variations on a fairly simple theme. Probably the other phyla shared common ancestors that were not cnidarian but may have been similar in many ways to flatworms though not identical with any modern type of flatworm.

Flatworms (*Platyhelminthes*) are acoelomate and unsegmented; they lack a blood system. Thus, in structure, they represent admirable ancestors for animals with any sort of body cavity, with and without segmentation. In their embryology and some other features they show similarities to molluscs and to annelids; earlier we pointed out that the two latter show certain resemblances. But there are marked differences between annelids and molluscs and it seems likely that each of these has evolved separately from a flatworm-like ancestor. In snails (*Mollusca*) the coelom is relatively small but spaces of the blood system form a haemocoel that functions as skeleton. In earthworms (*Annelida*), the blood is confined to tubes and the coelom forms a series of cavities which, like the main body muscles, are segmented. Which came first, blood system, segmentation or the coelom? There is no accepted answer and plenty of room for argument. To sum up so far, we can construct a diagram:

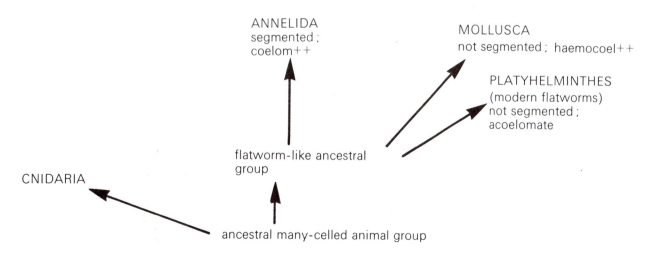

Arthropoda are segmented and coelomate and resemble annelids in certain ways. Adult arthropods have an extensive haemocoel and restricted coelom but early in development the coelom is much larger. Perhaps then the arthropod evolved from an ancestor resembling annelids by development of an exoskeleton and expansion of the haemocoel at the expense of the coelom. The possession of exoskeleton and jointed limbs appears to have been a major 'break-through' in body design judging by the great variety of arthropods and large number of species (see *IS*). Were these features evolved once only? Modern experts think it likely that they evolved at least twice (perhaps more often) and that the phylum *Arthropoda* is not a 'natural' group but represents several parallel lines of evolution from annelid-like ancestors. Add this to the diagram above.

The four groups Turbellaria, Mollusca, Annelida and Arthropoda can thus be linked loosely into one super-phylum (often called *SPIRALIA* or *PROTERO-STOMIA*). What of the others? The *Nematoda*, as already suggested, seem to represent a line of evolution with a unique body cavity, the pseudocoel. Possibly they evolved from a flatworm-like ancestor but they have many peculiar features and are usually placed in an isolated phylum (with some small animals not mentioned here). **Proterostomia**

The remaining groups that we have studied here are all coelomate but only the vertebrates are segmented. The phylum *Hemichordata*, including *Cephalodiscus*, is sometimes associated with the class Vertebrata into the phylum *Chordata*, which includes unsegmented animals without backbones as well as the segmented vertebrates; they share possession of slits in the pharynx called 'gills'. The early development of hemichordates and the non-vertebrate chordates shows marked resemblances to that of echinoderms, and there are other similarities. So *Echino-dermata* and *Chordata* can be linked into a super-phylum (called *RADIALIA* or *DEUTEROSTOMIA*). Possibly these evolved from a flatworm-like ancestor **Deuterostomia** by formation of coelomic cavities providing the hydro-skeleton for a small sedentary or sessile animal feeding with tentacles rather as *Cephalodiscus* does today. From this sort of animal evolved larger, motile animals including the Echinodermata, with their peculiar pentamerous symmetry and unique tube feet and the vertebrates with their segmented muscles and skeletons of cartilage or bone.

We can show this as a diagram:

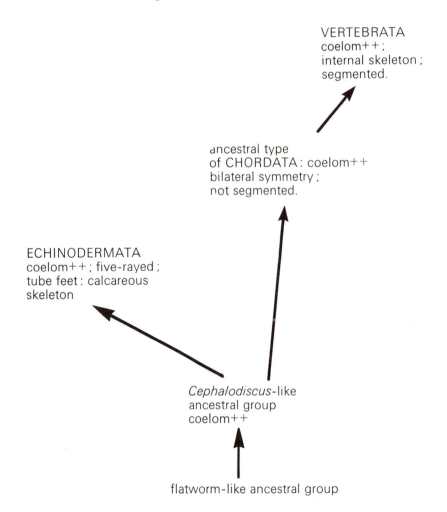

VERTEBRATA
coelom++;
internal skeleton;
segmented.

ancestral type
of CHORDATA: coelom++
bilateral symmetry;
not segmented.

ECHINODERMATA
coelom++; five-rayed;
tube feet: calcareous
skeleton

Cephalodiscus-like
ancestral group
coelom++

flatworm-like ancestral group

SUMMARIZE this attempt to make super-phyla by constructing a diagram including all the groups we have mentioned. Compare your version with that below.

We have constructed an 'evolutionary tree' or 'shrub', but you must realize that we have selected certain organisms and groups for study. Some that we have not mentioned have mosaics of features—some features like Radialia and others like Spiralia. These perhaps represent further separate lines of evolution. Our conclusions are based on circumstantial evidence and some of them are controversial. We shall never know the exact steps by which modern animals evolved. Is there any point then, in trying to unravel possible relationships?

Apart from the intellectual pleasure of the exercise, there is the practical point that it is easier to remember information if it fits into a clear frame of reference, such as that provided by a system of taxonomy and phylogeny. For this Course, it is important to understand how animals live today. But you must remember that function depends on structure and both structure and function have evolved as a result of mutation and natural selection. The 'options' open to any organsim are limited by its line of descent, by the genetic material it inherited from its parents; this can be traced back to its remote ancestors. Thus two organisms that live similar lives may solve their problems in very different ways because their genetic make-up is totally different and this means differences in structure and in biochemistry. In the last Unit of this Course we shall return to this theme of evolutionary relationships and review some examples of the evolution of physiological mechanisms.

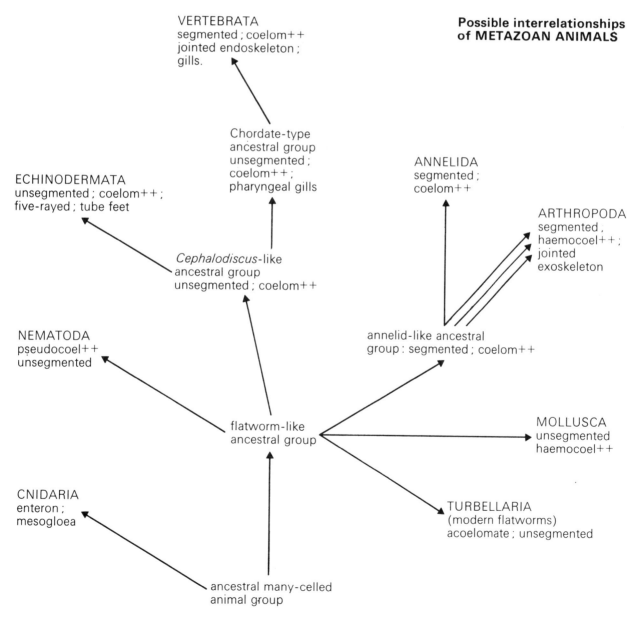

31

1.5 Summary

Comparative physiology is the study of functions in a variety of organisms. It must be based on some knowledge of the structure of various organisms.

The unit of classification is the *species*; problems in defining this are discussed in Section 1.1. Species may be associated into genera, these into families, these into orders, these into classes and the classes are associated into phyla (animals) or Divisions (plants).

Living organisms are divided among five kingdoms defined in Section 1.2. These are: viruses; bacteria and blue-green algae; fungi; plants; animals. Bacteria and blue-green algae are Procaryota; fungi, plants and animals are Eucaryota.

The Animal Kingdom is surveyed in Section 1.3 from the point of view of organization of the body in relation to movement. Various types of skeleton and arrangements of muscles are illustrated by examples, mostly of 'invertebrate' animals.

Grouping of phyla into larger groups, based on possible evolutionary relationships between them, makes it easier to organize in Section 1.4 the information given in Section 1.3. Since structure of living animals is the result of evolution from earlier forms, study of phylogeny may help to explain why similar functions are often carried out in different ways by distantly related animals living under similar conditions.

References

1   S100, Unit 19, *Evolution by Natural Selection*, Section 19.7.1.

2   S100, Unit 17, *The Genetic Code: Growth and Replication*, Section 17.1

3   S100, Unit 18, *Cells and Organisms*, Appendix 2 and Unit 21, *Unity and Diversity*, Section 21.7.

4   S100, Unit 18, Appendix 2.

Glossary

FISSION   A method of asexual reproduction when the organism divides its body into two parts that may be equal or unequal. Each part becomes a new organism and develops any missing organs or organelles.

METAZOANS   Many-celled animals. In some classifications, the Animal Kingdom is sub-divided into Protozoa (single-celled or acellular animals), Parazoa (sponges) and Metazoa.

PARTHENOGENESIS   Asexual reproduction by means of unfertilized eggs that develop into adults. The eggs may be produced by meiosis (and the resulting adults are then haploid) or there may be no meiosis (and the adults are then diploid and identical genetically with the mother). Aphids and honey bees reproduce parthenogenetically and also sexually.

PLASTIDS   Membrane-bound organelles of characteristic structure within cells. Those with green pigment are called chloroplasts but there are colourless plastids and some of other colours. They are produced by division of bodies called proplastids.

VEGETATIVE REPRODUCTION   Types of reproduction which do not involve a sexual process. These are common among plants, e.g. formation of bulbs and suckers. The daughter plants are identical genetically with the parent.

Self-assessment Questions

After Section 1.1

SAQ 1 (*Objective 1*)

The brown trout is *Salmo trutta* L.; the speckled trout is *Salvelinus fontinalis* (Mitchill) the Atlantic salmon is *Salmo salar* L; the char is *Salvelinus alpinus* (L); the grayling is *Thymallus thymallus* (L). All these fish belong to the family Salmonidae.

Mark the following statements as either TRUE or FALSE:

1 All five species of fish belong to the same genus.
2 All five species of fish belong to the same order.
3 All five species of fish belong to the same class.
4 All five species of fish belong to the same phylum.
5 The brown trout is more closely related to the speckled trout than it is to the Atlantic salmon.
6 The speckled trout is more closely related to the char than it is to the brown trout.
7 There is not enough information given to deduce whether the grayling is more closely related to the brown trout or to the char.

After Section 1.2

SAQ 2 (*Objective 4*)

In the following table, put + if the feature is found in the group; put 0 if it is not found.

Feature	PROCARYOTA	EUCARYOTA
1 Cell wall present, with cytoplasmic membrane of two layers visible under electron microscope inside it.		
2 Chlorophyll is found in lamellae in organelles called plastids.		
3 Mitochondria are present.		
4 Photosynthetic pigment is on lamellae that are not separated from the cytoplasm by membranes.		
5 Nuclei can divide by mitosis.		
6 Respiratory enzymes are concentrated within the cytoplasmic membrane but not in separate organelles.		
7 Nuclear material is separated from cytoplasm by unit membrane.		

After Section 1.3 or the whole Unit

SAQ 3 (*Objective 2(a)*)

Consider whether or not the following features of structure or life history are generally associated with either aquatic or terrestrial habits of life.

Mark A against those features generally associated with the aquatic habit of life.

Mark T against those features generally associated with the terrestrial habit of life.

Mark X against those features that are not specially associated with either aquatic or terrestrial habits of life.

(a) Wings are present [   ]; (b) the surface of the body is ciliated [   ]; (c) large eggs are laid that have a water-proofed shell [   ]; (d) very small eggs are laid that hatch into small larvae covered with cilia [   ]; (e) a blood-vascular system is present [   ]; (f) a coelom is present [   ]; (g) organs that function in air-breathing are present [   ]; (h) a shell is present [   ]; (j) jointed limbs are present [   ]; (k) tube feet are present [   ].

SAQ 4 (*Objective 2(b)*)

Fanworms, such as *Sabella* live in tubes.
Jelly-fish float in the upper waters of the oceans.
Oysters cement themselves to rocks.
Cockles burrow in muddy sand.
Coral polyps have a skeleton that is cemented to rocks.
Anemones live on rocks in tidal pools.
Flatworms glide over stones in streams.
Snails crawl over plants.
Locusts may fly in swarms.

Classify the animals mentioned above as:

A sedentary; B sessile; C motile.

After Section 1.3

SAQ 5 (*Objective 5*)

From the lists (A1–D4) below, select those statements that are typical of each of the groups: 1 Cnidaria; 2 turbellarians; 3 nematodes; 4 Annelida; 5 Mollusca; 6 Arthropoda; 7 Echinodermata; 8 vertebrates.

A1 A mouth is present but no anus.
A2 A mouth and anus are both present.
A3 A haemocoel is present.
A4 A pseudocoel is present.

B1 The muscle fibres are attached along their whole lengths to membrane.
B2 Longitudinal and circular muscles are present.
B3 Longitudinal muscles are present but no circular muscles.

C1 A coelom and a blood system are both present.
C2 A coelom is present but no blood-vascular system.
C3 The body is not segmented.
C4 Tube feet are present.

D1 A hydro-skeleton is present.
D2 Jointed limbs are present.
D3 A shell and a head-foot are both present.
D4 An endoskeleton is present.

SAQ 6 (*Objective 6*)

From the lists of information (A1–D4) below, select sets of organisms that illustrate: (a) convergence; and (b) adaptive radiation,

A1 Slugs creep by muscular waves passing along the foot.

A2 Fanworms live in tubes from which they push out a crown of ciliated tentacles.

A3 Leeches swim by dorso-ventral bending of the body.

A4 Crabs scuttle under rocks by sideways movements of their jointed legs.

B1 Jelly-fish swim by propelling water from under their 'umbrellas'.

B2 Spiders run across webs on jointed legs.

B3 Flatworms move by muscular waves passing along the ventral surface of their bodies.

B4 Water beetles swim by movements of their jointed legs.

C1 Grasshoppers and fleas leap high with their jointed legs.

C2 Fly maggots move by peristalsis.

C3 Cockles burrow in sand by movements of the foot.

C4 Sea-urchins crawl over rocks with tube feet.

D1 Earthworms burrow by peristalsis.

D2 Millipedes push their way into leaf litter on jointed legs.

D3 Some starfish burrow in sand with tube feet.

D4 Squids swim by propelling water from the mantle cavity through the siphon.

Self-assessment Answers and Comments

SAQ 1 1 False—three genera are listed: *Salmo, Salvelinus* and *Thymallus*.
2 True—since they all belong to the same family, they must all belong to the same order because an order is a taxon of higher rank than a family. Refer back to 1.1.1 if you are in doubt about this.
3 True—a class is a taxon of higher rank than family or order.
4 True—a phylum is a taxon of higher rank than family, order or class.
5 False—The brown trout and salmon both belong to the genus *Salmo* so they should be more closely related to each other than either is to the speckled trout, which belongs to the genus *Salvelinus*.
6 True—the char and speckled trout both belong to the genus *Salvelinus* whereas the brown trout belongs to the genus *Salmo*.
7 True—the three fish belong to different genera but to the same family. There is no indication of how the three genera are related within the family Salmonidae.

SAQ 2

Feature	Procaryota	Eucaryota
1	+	+
2	−	+
3	−	+
4	+	−
5	−	+
6	+	−
7	−	+

Procaryotes differ from eucaryotes in lacking membrane-bound organelles (features 2, 3 and 7 are absent but 4 and 6 are present). Because they lack a distinct nucleus, procaryotes do not undergo mitosis (feature 5). Both groups show feature 1. Refer back to early in Section 1.2 and to S100[4] for further information.

SAQ 3 (a) T—but you will be able to think of exceptions to the general rule that wings are of use only to terrestrial animals (penguins, water beetles).
(b) A—cilia only operate in a liquid medium.
(c) T—but again there are exceptions (insect eggs in water).
(d) A—small delicate eggs and larvae can survive only in water.
(e) X—the presence or absence of a blood-vascular system is more closely related to size, shape and activity of animals than to the medium in which they live. You will read about blood systems in Unit 5.
(f) X—the presence or absence of a coelom is a taxonomic character not related to the medium in which animals live.
(g) T—but there are aquatic animals that breathe air, as you will read in Unit 6.
(h) X—shells may serve a protective function on land and in water; molluscs typically have shells and the majority live in water whereas tortoises have shells and live on land while turtles are mainly aquatic.
(j) X—jointed limbs are typical of arthropods that may live either in water or on land; terrestrial groups of vertebrates also have jointed limbs (but of a different type). The possession of jointed limbs makes very active life on land possible.
(k) A—these structures can support a fairly heavy body only in water.

SAQ 4 A—*Sabella* and sea anemones are sedentary—they can move about but normally remain in one place (in a tube or on a rock).
B—oysters and coral polyps are sessile—they cannot move about once they have become fixed to rocks.
C—jelly-fish, cockles, flatworms, snails and locusts are all motile animals that spend much of their lives moving about.

SAQ 5 1 Cnidaria have: A1; B1; B2; C3; D1 (enteron cavity).
2 Turbellarians have: A1; B2; C3; you may have added D1, assuming that the parenchyma and spaces between the cells form a hydro-skeleton—it is debatable whether this really fits the definition.
3 Nematodes have: A2; A4; B3; C3; D1 (pseudocoel).
4 Annelids have: A2; B2; C1; D1 (coelomic cavities).
5 Molluscs have: A2; A3; C1; C3; D1 (haemocoel); D3. You may have hesitated between B2 and B3 and then chosen either of these. Molluscs have such a specialized muscular system that it is difficult to fit into these statements so the wisest course is probably to leave out both statements.
6 Arthropods have: A2; A3; C1; D2. Again the muscular system is specialized but some arthropods certainly have circular and longitudinal muscles so B2 can be added to the list. Some arthropods also have a hydro-skeleton (haemocoel) but they are in a minority so you should not have written down D1.
7 Echinoderms have: A2; B2; C2; C3; C4; D1 (coelomic cavities); D4. The hydro-skeleton is involved in the operation of the tube-feet.
8 Vertebrates have: A2; B2; C1; D2 (in some vertebrates only); D4.

If you are in doubt about any of these features, read Section 1.2 of the Unit and look at the appropriate pages of *IS*.

SAQ 6 (a) Similarities between organisms that are unrelated but have similar habits are examples of convergence. There is convergence between: A1 and B3 in the way they crawl on a flat muscular foot; B1 and D4 in the way they swim by jet propulsion; C2 and D1 in the way they crawl by peristaltic contractions of the muscles. These are the clearest examples of convergence in this matrix—comparisons between, for example, C3 and D3 or between C3 and C2+D1 are too far-fetched to be allowed.
(b) Adaptive radiation is the term used when related animals are adapted to different habits of life. In the matrix, the following sets illustrate adaptive radiation:
annelids—A2, A3 and D1; arthropods A4, B2, B4, C1, D2 (all these are comparable as adults; C2 illustrates another way of life but you should argue that it is not legitimate to compare larvae (maggots) with adults); molluscs—A1, C3, and D4; echinoderms—C4 and D3.

Answers to In-text Questions

ITQ 1 As long as the volume of incompressible liquid in the enteron cavity is constant (i.e. as long as the mouth is closed) contraction of one set of muscles must result in the stretching of the other set of muscles. You can imitate this situation with a rubber balloon—inflate it, then squeeze it in one place—note that there is stretching of the rubber in some other place and the volume of air remains constant.

ITQ 2 The mesogloea is deformed by contraction of the muscles—when they relax, it regains its original shape and at the same time stretches the muscle fibres. The animal swims by alternate contractions and relaxations of the circular muscles.

ITQ 3 The situation is more complicated than in a flatworm: the shape of the body is maintained by the tough, fibrous cuticle on the outside of the worm exerting a high pressure on the liquid in the pseudocoel. When a muscle contracts, distortion of shape is opposed by the cuticle and the pressure in the pseudocoel is slightly increased, tending to stretch the muscles on the other side of the body; on relaxation, the muscle is quickly stretched as the pseudocoel pressure restores the round body shape.

ITQ 4 There are three sets of muscles: an outer layer of circular muscles, inner blocks of longitudinal muscles and muscles in the outer wall of the gut (causing gut movements). There is a gut cavity and there are large cavities between the muscles of the gut wall and the longitudinal muscles. There are also other tubular structures connected with excretion and gonads, and there is a system of tubes carrying blood—the blood-vascular system. The large cavities between the gut wall and body wall form the *coelom*; it is filled with coelomic fluid.

ITQ 5 (a) The segment will change shape, becoming shorter but wider; the circular muscles will be stretched (as in Fig. 17(a)).
(b) The segment will change shape becoming longer and thinner; the longitudinal muscles will be stretched (as in Fig. 17(b)).

ITQ 6 (a) Vertical bending.
(b) The longitudinal muscles.
(c) The dorsal longitudinal muscles are antagonists of the ventral longitudinal muscles.
(d) The botryoidal tissue acts as the skeleton.

ITQ 7 The tube feet are hollow; their cavities are a part of a series of coelomic tubes and cavities called the 'water-vascular system'. Each ampulla has sets of muscles arranged diagonally at right angles to each other; when these contract fluid from the bladder is forced into the cavity of the tube foot and the muscles are stretched. When the muscles of the foot contract, fluid is forced back into the bladder. This seems the obvious explanation from the information given here, but it is not the whole story, as you will learn from the TV programme of this Unit.

ITQ 8 No. The arthropod skeleton is outside the muscles (hence called an *exo*skeleton) whereas the vertebrate skeleton is within the musculature (hence called an *endo*skeleton). The skeletons are also composed of different substances: chitin in arthropods, with either calcium carbonate or protein added to it; cartilage or bone in vertebrates.

ITQ 9 No. In arthropods, the main body cavity is the haemocoel whereas in vertebrates it is the coelom.

ITQ 10 No; no. In arthropods, the nerve cord lies below (ventral to) the gut whereas in vertebrates it is above (dorsal to) the gut. The arthropod nerve cord is solid and double in structure whereas that of vertebrates is a hollow tube.

ITQ 11 No. If the skeleton is sufficiently rigid, the antagonistic muscles can be arranged so that contraction of one set will stretch another set as in Figure 28.

ITQ 12 The internal skeleton of fishes and mammals provides the framework against which the muscles can contract, stretching their antagonists; fluid pressure in the coelom plays no part in this system.

Table 2

	Flatworm	Roundworm	Earthworm	Nereis	Sabella	Leech
Is a gut present?	√	√	√	√	√	√
Has the gut muscles in its walls?	×	×	√	√	√	√
Is a pseudocoel present?	×	√	×	×	×	×
Is a coelom present? (a) filled with fluid (b) filled with cells	×	×	(a) √	(a) √	(a) √	(b) √
Are circular body muscles present?	√	×	√	√	√	√
Are longitudinal muscles present?	√	√	√	√	√	√
Which muscles are involved in main movement of body?	Longitudinal	Longitudinal	Longitudinal & Circular	Longitudinal	Longitudinal & Circular	Longitudinal
Are blood vessels present?	×	×	√	√	√	√

Table 3

	Column 1 Coelomic fluid (including cavity of gonad)	Column 2 Fluid in haemocoel or blood vessels
Earthworm	B C	A
Snail	B	A C

Table 4

Phylum or Class	CNIDARIA	TURBELLARIA	NEMATODA	ANNELIDA	MOLLUSCA	HEMICHORDATA	ECHINODERMATA	ARTHROPODA	VERTEBRATA
Animal	Polyp	Flatworm	Roundworm	Earthworm	Snail	*Cephalodiscus*	Starfish	Insect	Mammal
Gut cavity (or enteron)	++	++	+	+	+	+	+	+	+
Pseudocoel	0	0	++	0	0	0	0	0	0
Coelom	0	0	0	++	+	++	++	+	++
Blood system (or haemocoel)	0	0	0	+	++	+	0	++	+
Segmented muscles	0	0	0	+	0	0	0	+	+

COMPARATIVE PHYSIOLOGY

1 Introduction—The Animal Kingdom
2 Plant Structure, Diversity and Habit
3 Movement of Water and Organic Substances in Plants
4 Nutrition, Feeding Mechanisms and Digestion in Animals
5 Circulation of the Blood
6 Respiratory Mechanisms
7 Plant Growth and Differentiation
8 Hormones and the Reproductive Cycle
9 Hormones and Homeostasis: Blood Sugar and Blood Calcium
10 Hormones and Homeostasis: Osmoregulation and Excretion
11 Physiological Mechanisms and Physiological Evolution